~A BINGO BOOK~

Geometry and Measurement Bingo Book

COMPLETE BINGO GAME IN A BOOK

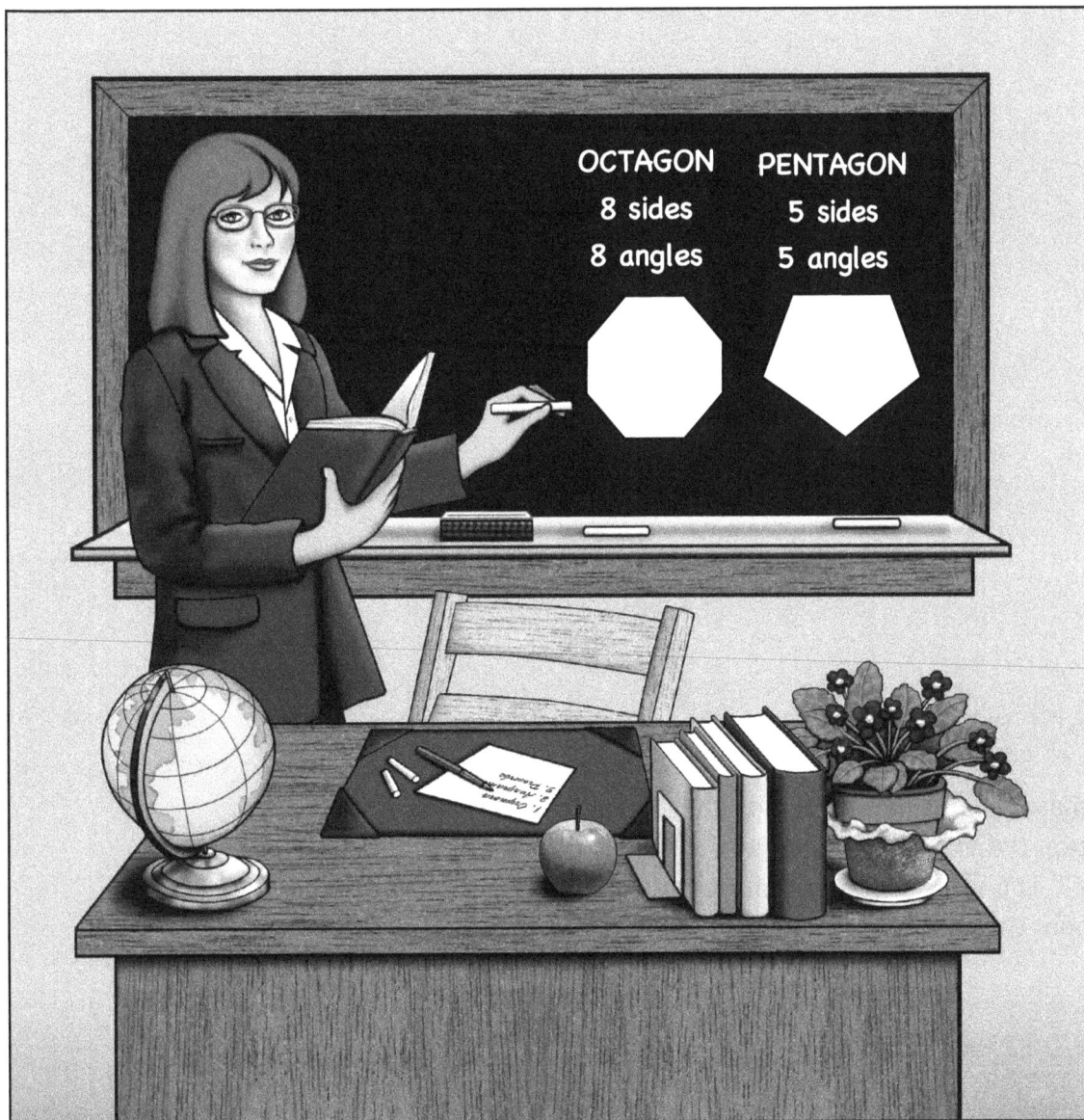

OCTAGON
8 sides
8 angles

PENTAGON
5 sides
5 angles

Written By Rebecca Stark

Educational Books 'n' Bingo

TITLE: Geometry and Measurement Bingo
AUTHOR: Rebecca Stark

ISBN 978-0-87386-453-4

Educational Books 'n' Bingo

Printed in the U.S.A.

GEOMETRY AND MEASUREMENT BINGO DIRECTIONS

INCLUDED:

List of Terms

Templates for Additional Terms and Clues

2 Clues per Term

30 Unique Bingo Cards

Markers

1. **Either cut apart the book or make copies of ALL the sheets. You might want to make an extra copy of the clue sheets to use for introduction and review. Keep the sheets in an envelope for easy reuse.**

2. Cut apart the call cards with terms and clues.

3. Pass out one bingo card per student. There are enough for a class of 30.

4. Pass out markers. You may cut apart the markers included in this book or use any other small items of your choice.

5. Decide whether or not you will require the entire card to be filled. Requiring the entire card to be filled provides a better review. However, if you have a short time to fill, you may prefer to have them do the just the border or some other format. Tell the class before you begin what is required.

6. There are 50 terms. Read the list before you begin. If there are any terms that have not been covered in class, you may want to read to the students the term and clues before you begin.

7. There is a blank space in the middle of each card. You can instruct the students to use it as a free space or you can write in answers to cover terms not included. Of course, in this case you would create your own clues. (Templates provided.)

8. Shuffle the cards and place them in a pile. Two or three clues are provided for each term. If you plan to play the game with the same group more than once, you might want to choose a different clue for each game. If not, you may choose to use more than one clue.

9. Be sure to keep the cards you have used for the present game in a separate pile. When a student calls, "Bingo," he or she will have to verify that the correct answers are on his or her card AND that the markers were placed in response to the proper questions. Pull out the cards that are on the student's card keeping them in the order they were used in the game. Read each clue as it was given and ask the student to identify the correct answer from his or her card.

10. If the student has the correct answers on the card AND has shown that they were marked in response to the *correct questions,* then that student is the winner and the game is over. If the student does not have the correct answers on the card OR he or she marked the answers in response to *the wrong questions,* then the game continues until there is a proper winner.

11. If you want to play again, reshuffle the cards and begin again.

Have fun!

TERMS INCLUDED

ACUTE ANGLE

ACUTE TRIANGLE

ANGLE

AREA

CAPACITY

CELSIUS

CHORD

CIRCLE

CIRCUMFERENCE

COMPLEMENTARY

CONGRUENT

CUBE

DIAMETER

EQUILATERAL TRIANGLE

FAHRENHEIT

GALLON

GEOMETRY

HEXAGON

INTERSECT

ISOSCELES TRIANGLE

LINE

LINEAR MEASUREMENT

METER

OBTUSE ANGLE

OCTAGON

PARALLEL

PARALLELOGRAM

PENTAGON

PERIMETER

PERPENDICULAR

PI

PLANE

POINT

POLYGON

PROTRACTOR

QUADRILATERAL

RADIUS

RAY

RECTANGLE

RIGHT ANGLE

RIGHT TRIANGLE

SCALENE TRIANGLE

SQUARE

STRAIGHT ANGLE

SUPPLEMENTARY

TIME

TRAPEZOID

TRIANGLE

VOLUME

WEIGHT

Additional Terms

Choose as many additional terms as you would like and write them in the squares. Repeat each as desired.
Cut out the squares and randomly distribute them to the class.
Instruct the students to place their square on the center space of their card.

Geometry and Measurement Bingo

Clues for
Additional Terms

Write three clues for each of your additional terms.

_____

1.

2.

3.

_____

1.

2.

3.

_____

1.

2.

3.

_____

1.

2.

3.

_____

1.

2.

3.

_____

1.

2.

3.

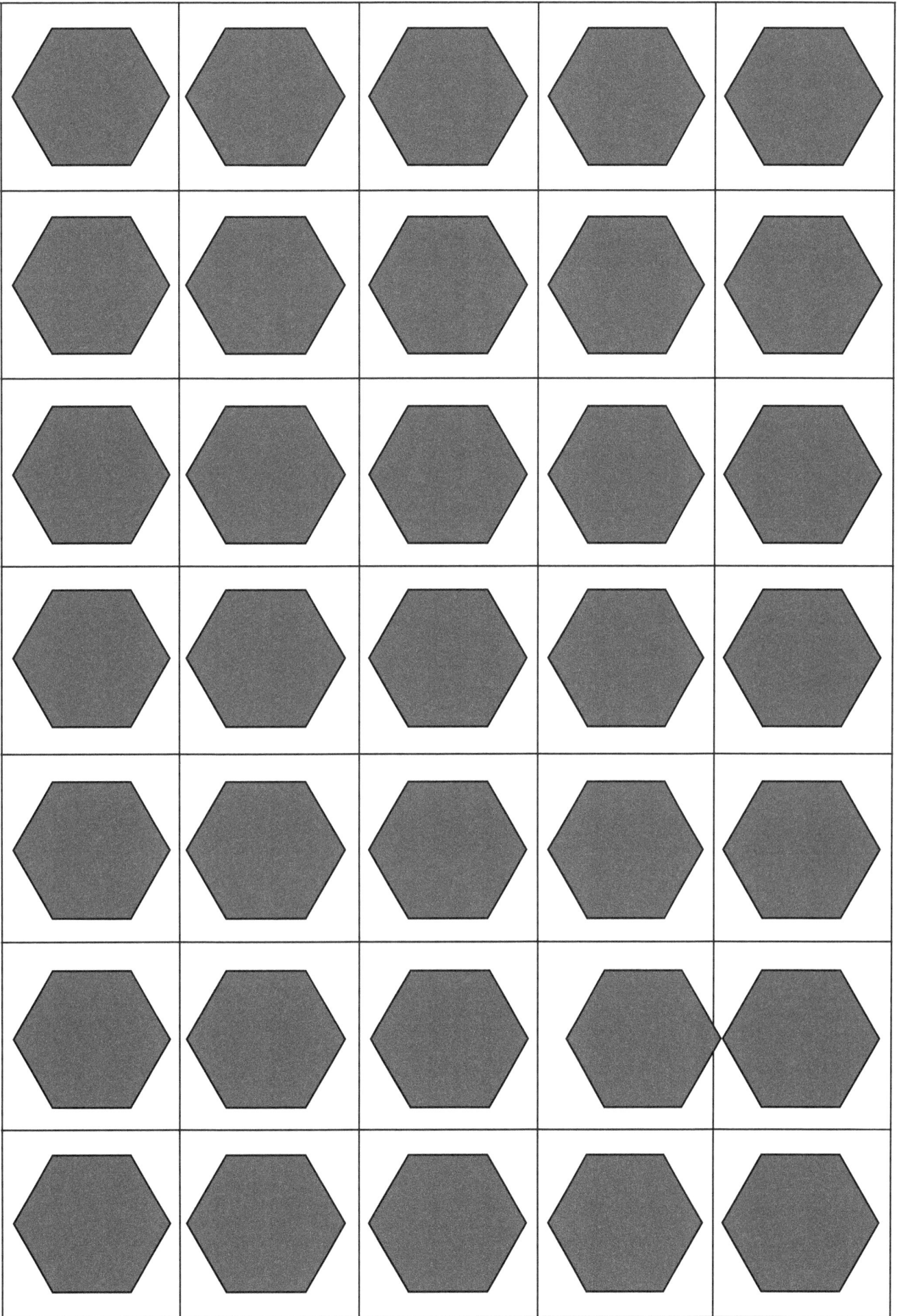

Acute Angle 1. It measures less than 90°. 2. It is smaller than a right angle. 3. A 60° angle is one.	**Acute Triangle** 1. This polygon has 3 acute angles. 2. If all 3 of a triangle's angles are unequal and less than 90°, it is this kind of triangle. 3. An equilateral triangle is also this kind of triangle.
Angle 1. One is formed when 2 rays have a common endpoint. 2. Its common endpoint is called a vertex. 3. We measure it in degrees.	**Area** 1. The measure of square units within a figure. 2. You get the ___ of a rectangle by multiplying its length times its width. 3. You get the ___ of a triangle by using the formula $A = 1/2\ bh$, where b is the base and h is the height.
Capacity 1. Amount of liquid a container can hold. 2. Pints, quarts and gallons are U.S. customary units of ___. 3. Milliliters, and liters are metric units of ___.	**Celsius** 1. The Metric System uses this scale to measure temperature. 2. On this scale, water boils at 100°. 3. On this scale, water freezes at 0°.
Chord 1. Any line segment that joins two points on a circle. 2. A diameter is one, but a radius is not. 3. If a ___ goes through the center point of a circle, we call it a diameter.	**Circle** 1. A closed curve made up of all the points at the same distance from a given point is one. 2. A v-shaped instrument called a compass is sometimes used to draw one. 3. The center of this closed curve is a point, which can be used to name the figure.
Circumference 1. The distance around a circle. 2. The name for the perimeter of a circle. 3. To find it, multiply π times the diameter.	**Complementary** 1. A pair of angles that equal 90° when combined are said to be this. 2. A 75° angle and a 15° angle are this. 3. A 60° angle and a 30° angle are this.

Congruent	**Cube**
1. Figures with the same size and shape are this. 2. ___ figures would coincide exactly if one were placed on top of the other. 3. If all 3 sides in two triangles are the same size, the triangles are ___.	1. This figure has 6 congruent faces. 2. All 6 of this figure's faces are square. 3. If each of its sides is 3 inches, its volume is 27 cubic inches.
Diameter	**Equilateral Triangle**
1. A chord that passes through the center of a circle. 2. It equals two radii. 3. To find the circumference of a circle, we multiply this times π.	1. This figure has 3 equal sides and 3 equal angles. 2. Each of its 3 angles measures 60°. 3. A triangle with three 3-inch sides is this.
Fahrenheit	**Gallon**
1. The U.S. Customary System uses this scale to measure temperature. 2. On this scale, water boils at 212°. 3. On this scale, water freezes at 32°.	1. 4 quarts 2. 8 pints 3. 128 ounces
Geometry	**Hexagon**
1. Branch of mathematics that deals with points, lines, angles and surfaces. 2. Branch of mathematics concerned with the size, shape and relative positions of figures. 3. Study of 2- and 3-dimensional figures.	1. This polygon has 6 sides. 2. This polygon has 6 interior angles. 3. This polygon has one more side than a pentagon.
Intersect	**Isosceles Triangle**
1. To cross each other at a point. 2. What perpendicular lines do at right angles. 3. What a pair of parallel lines never do.	1. This kind of triangle has 2 equal sides and 2 equal angles. 2. A triangle with two 65° angles is one. 3. A triangle with the angles 40°, 40° and 100° is an obtuse ___ triangle.

Geometry and Measurement Bingo

Line
1. A set of points along a straight path.
2. A ___ segment includes all the points that fall between two endpoints.
3. Arrows pointing in both directions written over 2 points signifies a ___.

Linear Measurement
1. The measure of the length of an object.
2. Inches, feet, yards and miles are U.S. customary units of ___.
3. Meters, centimeters and kilometers are metric units of ___.

Meter
1. 1,000 millimeters
2. 100 centimeters
3. 10 decimeters

Obtuse Angle
1. It measures more than 90° but less than 180°.
2. A 120° angle is one.
3. A 175° angle is one.

Octagon
1. This polygon has 8 sides.
2. This polygon has 8 angles.
3. This polygon has twice as many sides as a quadrilateral.

Parallel
1. Lines that are in the same plane but never meet are called ___ lines.
2. ___ lines are at the same distance apart at all points.
3. These lines never intersect.

Parallelogram
1. A quadrilateral whose opposite sides are parallel is one.
2. To find its area, use the formula *A = bh,* or *Area + base x height.*
3. A square is one; so is a rhombus.

Pentagon
1. A 5-sided polygon.
2. This polygon has 5 interior angles.
3. This polygon has more sides than a quadrilateral but less sides than a hexagon.

Perimeter
1. The distance around a geometric shape.
2. The ___ of a polygon is found by adding all of its sides.
3. If one side of a square is 3 inches, its ___ is 12 inches.

Perpendicular
1. Lines that intersect at right angles are called ___ lines.
2. When ___ lines intersect, they form square corners at the point where they meet.
3. Four right angles are formed by the intersection of lines of this type.

Pi (π) 1. To find the circumference of a circle, multiply this times the diameter. 2. To find the area of a circle, multiply this times the radius squared. 3. It represents the ratio of the circumference of a circle to its diameter; its value is about 3.1459.	**Plane** 1. A 2-dimensional surface. 2. The set of all the points on an endless flat surface. 3. A flat surface that extends into infinity in all directions.
Point 1. An exact location in space. 2. In the line segment *BC, B* is an end___; so is *C.* 3. In the ray *BC, B* is an end___ and *C* is a ___ on the line.	**Polygon** 1. A closed figure made up of 3 or more line segments and the same number of angles. 2. It is named according to the number of sides it has. For example, a 3-sided one is called a triangle. 3. Triangles, quadrilaterals, hexagons and octagons are all examples of this.
Protractor 1. An instrument used to measure and draw angles. 2. These instruments are either circular or semicircular. 3. A semicircular one is divided into degrees from 0° to 180°.	**Quadrilateral** 1. Any 4-sided polygon. 2. A parallelogram is this type of polygon; so is a trapezoid. 3. The sum of its 4 interior angles is 360°.
Radius 1. Any line segment with one endpoint at the center of the circle and the other on the circle. 2. It equals one-half of the diameter. 3. A diameter is really two of these.	**Ray** 1. A segment of a line that extends from 1 endpoint and continues indefinitely in the other direction. 2. You name it by is endpoint and a point on the line. 3. The letters BC under an arrow pointing right is called ___ BC.
Rectangle 1. Like a square and a rhombus, this quadrilateral is a parallelogram. 2. A parallelogram with 2 sets of equal sides and 4 equal angles is called a ___. 3. If this parallelogram has two 3-inch sides and two 4 inch sides, its area would be 12 square inches.	**Right Angle** 1. An angle that measures 90°. 2. It is larger than an acute angle and smaller than an obtuse angle. 3. This kind of angle forms a square corner.

Geometry and Measurement Bingo

Right Triangle
1. This kind of triangle has a right angle.
2. Two of its 3 angles add up to 90°.
3. It has 3 sides and 3 angles; one of its angles is 90°.

Scalene Triangle
1. If all 3 sides of a triangle are a different length, it is this kind of triangle.
2. A triangle with these lengths is one: 3 cm, 4 cm and 5 cm.
3. A triangle with 2 equal sides is not one.

Square
1. We call a parallelogram this when it has equal sides and equal angles.
2. Each of its 4 angles is 90°. Its sides are equal in length.
3. A figure with four sides measuring 5 inches and four equal angles is one.

Straight Angle
1. It measures 180°.
2. It is larger than an obtuse angle.
3. It twice as large as a right angle.

Supplementary
1. A pair of angles that equal 180° when combined are said to be this.
2. A 160° angle and a 20° angle are this.
3. A 125° angle and a 55° angle are this.

Time
1. Minute, hour, day & week are units of ___.
2. To change from a smaller unit of ___ to a larger one—such as from weeks to years— you divide.
3. To change from a larger unit of ___ to a smaller one—such as from years to weeks—you multiply.

Trapezoid
1. A quadrilateral with only two parallel sides.
2. It is like a parallelogram in that it has 4 sides. It is unlike a parallelogram because only 2 of its sides are parallel.
3. To find the area of this kind of quadrilateral, use the formula $A = 1/2\ (b_1 + b_2)$. (*b* refers to base)

Triangle
1. This polygon has 3 sides and 3 angles.
2. Its 3 interior angles add up to 180°.
3. To name it, use its 3 vertices.

Volume
1. The measure of the amount of space in a container.
2. Formula for the ___ of a cube is $V = s^3$, or side cubed.
3. Multiply length x width x height to find the ___ of a rectangular 3-dimensional figure.

Weight
1. The heaviness of an object.
2. Ounces and pounds are U.S. customary units of ___.
3. Grams and kilograms are metric units of ___.

Geometry and Measurement Bingo

Geometry and Measurement Bingo

Linear Measurement	Acute Angle	Capacity	Octagon	Parallelogram
Equilateral Triangle	Diameter	Scalene Triangle	Meter	Triangle
Area	Straight Angle		Square	Perimeter
Celsius	Gallon	Volume	Polygon	Obtuse Angle
Ray	Intersect	Complementary	Pi	Perpendicular

Geometry and Measurement Bingo

Celsius	Area	Point	Trapezoid	Fahrenheit
Obtuse Angle	Meter	Circle	Gallon	Radius
Cube	Intersect		Geometry	Volume
Rectangle	Right Triangle	Straight Angle	Right Angle	Perpendicular
Triangle	Scalene Triangle	Complementary	Equilateral Triangle	Pi

Geometry and Measurement Bingo

Celsius	Volume	Meter	Polygon	Area
Intersect	Acute Triangle	Congruent	Acute Angle	Pentagon
Gallon	Scalene Triangle		Radius	Angle
Straight Angle	Cube	Ray	Rectangle	Point
Pi	Equilateral Triangle	Complementary	Right Angle	Geometry

Geometry and Measurement Bingo

Straight Angle	Radius	Capacity	Equilateral Triangle	Geometry
Plane	Chord	Acute Angle	Trapezoid	Area
Square	Rectangle		Parallelogram	Octagon
Volume	Hexagon	Scalene Triangle	Complementary	Circle
Pentagon	Triangle	Weight	Pi	Perimeter

Geometry and Measurement Bingo

Triangle	Parallelogram	Gallon	Circle	Equilateral Triangle
Plane	Volume	Congruent	Fahrenheit	Acute Triangle
Capacity	Perimeter		Protractor	Line
Perpendicular	Geometry	Linear Measurement	Right Angle	Diameter
Meter	Complementary	Area	Straight Angle	Square

Geometry and Measurement Bingo

Angle	Radius	Point	Geometry	Perimeter
Polygon	Gallon	Diameter	Acute Angle	Area
Trapezoid	Parallel		Chord	Fahrenheit
Complementary	Ray	Right Angle	Weight	Capacity
Obtuse Angle	Volume	Linear Measurement	Square	Hexagon

Geometry and Measurement Bingo

Linear Measurement	Radius	Line	Protractor	Meter
Parallel	Fahrenheit	Intersect	Circumference	Plane
Point	Octagon		Geometry	Chord
Straight Angle	Rectangle	Congruent	Celsius	Cube
Complementary	Equilateral Triangle	Right Angle	Weight	Angle

Geometry and Measurement Bingo: Card No. 7

Geometry and Measurement Bingo

Square	Radius	Circumference	Polygon	Chord
Plane	Capacity	Trapezoid	Perimeter	Circle
Hexagon	Time		Geometry	Parallelogram
Pi	Straight Angle	Celsius	Parallel	Rectangle
Scalene Triangle	Complementary	Weight	Gallon	Obtuse Angle

Geometry and Measurement Bingo

Geometry	Meter	Intersect	Hexagon	Equilateral Triangle
Parallel	Fahrenheit	Square	Gallon	Radius
Pentagon	Linear Measurement		Diameter	Acute Triangle
Acute Triangle	Perpendicular	Ray	Protractor	Line
Rectangle	Right Angle	Congruent	Celsius	Parallelogram

Geometry and Measurement Bingo: Card No. 9

Geometry and Measurement Bingo

Celsius	Polygon	Chord	Trapezoid	Hexagon
Perimeter	Circle	Acute Angle	Diameter	Geometry
Time	Radius		Octagon	Cube
Ray	Perpendicular	Acute Triangle	Right Angle	Pentagon
Congruent	Parallel	Point	Triangle	Square

Geometry and Measurement Bingo

Angle	Radius	Gallon	Diameter	Parallel
Circumference	Pentagon	Protractor	Fahrenheit	Acute Angle
Plane	Geometry		Point	Intersect
Congruent	Area	Right Angle	Equilateral Triangle	Celsius
Obtuse Angle	Complementary	Linear Measurement	Weight	Meter

Geometry and Measurement Bingo

Meter	Parallelogram	Complementary	Polygon	Geometry
Intersect	Obtuse Angle	Capacity	Weight	Diameter
Linear Measurement	Line		Perimeter	Trapezoid
Pentagon	Rectangle	Fahrenheit	Celsius	Plane
Radius	Circumference	Time	Parallel	Circle

Geometry and Measurement Bingo

Diameter	Parallelogram	Angle	Straight Angle	Perimeter
Capacity	Circumference	Fahrenheit	Geometry	Cube
Polygon	Circle		Intersect	Line
Square	Right Angle	Chord	Time	Celsius
Complementary	Perpendicular	Weight	Linear Measurement	Protractor

Geometry and Measurement Bingo: Card No. 13

Geometry and Measurement Bingo

Equilateral Triangle	Fahrenheit	Gallon	Geometry	Obtuse Angle
Circle	Linear Measurement	Pentagon	Diameter	Radius
Volume	Octagon		Point	Congruent
Perpendicular	Right Angle	Time	Chord	Angle
Complementary	Trapezoid	Cube	Parallel	Square

Geometry and Measurement Bingo

Protractor	Fahrenheit	Gallon	Meter	Polygon
Angle	Point	Acute Angle	Capacity	Obtuse Angle
Perimeter	Linear Measurement		Area	Radius
Complementary	Pentagon	Circumference	Right Angle	Diameter
Parallel	Rectangle	Weight	Hexagon	Intersect

Geometry and Measurement Bingo

Chord	Pentagon	Circumference	Hexagon	Right Triangle
Trapezoid	Cube	Line	Plane	Octagon
Acute Triangle	Parallelogram		Perimeter	Intersect
Straight Angle	Circle	Complementary	Protractor	Celsius
Parallel	Supplementary	Weight	Rectangle	Radius

Geometry and Measurement Bingo

Congruent	Quadrilateral	Isosceles Triangle	Pentagon	Equilateral Triangle
Protractor	Parallel	Right Angle	Octagon	Line
Fahrenheit	Square		Supplementary	Circumference
Perpendicular	Obtuse Angle	Celsius	Gallon	Cube
Ray	Acute Triangle	Meter	Polygon	Parallelogram

Geometry and Measurement Bingo

Hexagon	Time	Circle	Diameter	Trapezoid
Radius	Congruent	Ray	Perimeter	Parallel
Fahrenheit	Cube		Isosceles Triangle	Capacity
Perpendicular	Acute Angle	Right Angle	Celsius	Point
Supplementary	Pentagon	Gallon	Quadrilateral	Angle

Geometry and Measurement Bingo

Perimeter	Angle	Pentagon	Circumference	Time
Protractor	Polygon	Radius	Meter	Octagon
Quadrilateral	Equilateral Triangle		Diameter	Area
Point	Supplementary	Ray	Rectangle	Isosceles Triangle
Capacity	Right Triangle	Parallel	Square	Weight

Geometry and Measurement Bingo

Time	Quadrilateral	Polygon	Pentagon	Weight
Circle	Intersect	Plane	Ray	Trapezoid
Parallelogram	Line		Straight Angle	Acute Angle
Triangle	Scalene Triangle	Pi	Rectangle	Supplementary
Volume	Square	Right Triangle	Celsius	Isosceles Triangle

Geometry and Measurement Bingo: Card No. 20

Geometry and Measurement Bingo

Protractor	Angle	Plane	Pentagon	Triangle
Parallelogram	Isosceles Triangle	Chord	Circumference	Linear Measurement
Cube	Parallel		Quadrilateral	Gallon
Ray	Meter	Supplementary	Perpendicular	Square
Straight Angle	Right Triangle	Weight	Congruent	Rectangle

Geometry and Measurement Bingo

Hexagon	Point	Isosceles Triangle	Capacity	Diameter
Trapezoid	Polygon	Area	Circumference	Acute Triangle
Circle	Octagon		Linear Measurement	Line
Supplementary	Perpendicular	Rectangle	Acute Angle	Equilateral Triangle
Right Triangle	Congruent	Quadrilateral	Cube	Plane

Geometry and Measurement Bingo

Chord	Quadrilateral	Meter	Capacity	Weight
Angle	Time	Parallel	Protractor	Acute Angle
Point	Diameter		Pi	Linear Measurement
Cube	Right Triangle	Supplementary	Congruent	Rectangle
Triangle	Scalene Triangle	Square	Ray	Isosceles Triangle

Geometry and Measurement Bingo

Chord	Time	Equilateral Triangle	Quadrilateral	Circumference
Isosceles Triangle	Weight	Plane	Trapezoid	Linear Measurement
Line	Hexagon		Diameter	Cube
Triangle	Pi	Supplementary	Congruent	Parallelogram
Volume	Straight Angle	Right Triangle	Polygon	Scalene Triangle

Geometry and Measurement Bingo

Straight Angle	Plane	Quadrilateral	Gallon	Isosceles Triangle
Acute Angle	Perpendicular	Protractor	Chord	Acute Triangle
Parallelogram	Circumference		Pi	Supplementary
Area	Triangle	Scalene Triangle	Right Triangle	Octagon
Weight	Equilateral Triangle	Circle	Parallel	Volume

Geometry and Measurement Bingo: Card No. 25

Geometry and Measurement Bingo

Isosceles Triangle	Quadrilateral	Point	Trapezoid	Hexagon
Ray	Polygon	Circumference	Time	Chord
Perpendicular	Pi		Octagon	Straight Angle
Congruent	Capacity	Triangle	Right Triangle	Supplementary
Line	Parallel	Gallon	Scalene Triangle	Volume

Geometry and Measurement Bingo

Point	Circle	Quadrilateral	Time	Intersect
Triangle	Pi	Protractor	Supplementary	Acute Triangle
Right Angle	Scalene Triangle		Right Triangle	Straight Angle
Hexagon	Angle	Plane	Volume	Acute Angle
Parallel	Octagon	Isosceles Triangle	Area	Line

Geometry and Measurement Bingo

Perimeter	Time	Area	Quadrilateral	Chord
Intersect	Isosceles Triangle	Pi	Trapezoid	Octagon
Scalene Triangle	Cube		Line	Ray
Celsius	Hexagon	Parallel	Right Triangle	Supplementary
Capacity	Fahrenheit	Obtuse Angle	Volume	Triangle

Geometry and Measurement Bingo

Isosceles Triangle	Time	Hexagon	Protractor	Fahrenheit
Perpendicular	Ray	Plane	Line	Area
Parallelogram	Pi		Perimeter	Quadrilateral
Intersect	Triangle	Geometry	Right Triangle	Supplementary
Chord	Acute Triangle	Volume	Angle	Scalene Triangle

Geometry and Measurement Bingo

Equilateral Triangle	Quadrilateral	Trapezoid	Geometry	Supplementary
Acute Angle	Time	Point	Octagon	Diameter
Perpendicular	Acute Triangle		Line	Plane
Volume	Angle	Capacity	Right Triangle	Pi
Triangle	Meter	Scalene Triangle	Isosceles Triangle	Area